WHAT IS CETOLOGY?

Cetology is the study of whales, dolphins, and porpoises. It explores their traits, **habitats,** biology, and interactions with the **environment**.

The scientists who study whales and dolphins are called **CETOLOGISTS.**

ARE WHALES AND DOLPHINS SMARTER THAN HUMANS?

DISCOVER THE SCIENCE BEHIND **CETOLOGY**
(see-TOL-uh-jee)

Written by Eliza Jeffery
Illustrated by Lays Bittencourt

Words that are tricky to understand are in **bold**. Find out what they mean in the glossary.

Words that are difficult to say are in *italics*. Find out how to say them at the back of the book.

Over seventy years ago, scientists studied a group of bottlenose dolphins and tracked all the noises they made. They found that these creatures were having conversations with each other!

"I think they're saying **hello!**"

Because of these early studies, we now have the science of cetology. By diving into the world's oceans, *cetologists* have been able to learn some pretty impressive things about these underwater **mammals**.

Cetologists were excited to find that whales and dolphins are great problem-solvers. And they are really smart when they hunt for food. One of their great tricks is using their **blowholes** to make a ring of bubbles. These bubble nets trap lots of fish in one place!

One dolphin **species** even uses tools when it looks for food! In Shark Bay, bottlenose dolphins are clever enough to use **sea sponges** to protect their noses from sharp objects when they search the ocean floor.

Another reason that whales and dolphins are so clever at finding food is because they are experts at communicating! Their language is passed down through families.

Beluga whales are known for their lengthy underwater chats, and some whales are clever enough to learn the language of dolphins too!

Animals need memory to learn, just like humans do. And whales have amazing memories! This helps them to **migrate** every year to important places like their feeding grounds.

Blue whales can live up to 100 years and still remember the routes they followed when they were young, and the other whales that swam with them.

Like us, whales and dolphins are social animals. Sperm whales live in groups called pods, which are led by females. These pods form bonds that last a lifetime! Sperm whales even babysit each other's young in these groups.

Maybe these smart abilities shouldn't surprise us – sperm whales have the largest brains of any animal in the world! Brain size and functions are ways cetologists can start to understand their

VERY SPECIAL ABILITIES...

HUMAN'S BRAIN
DOLPHIN'S BRAIN
SPERM WHALE'S BRAIN

And whales aren't just smart when they're awake. They are clever when they sleep too! Whales can't breathe underwater, so must come to the surface regularly for air. To be safe while they are sleeping, whales only rest one half of their brains at a time!

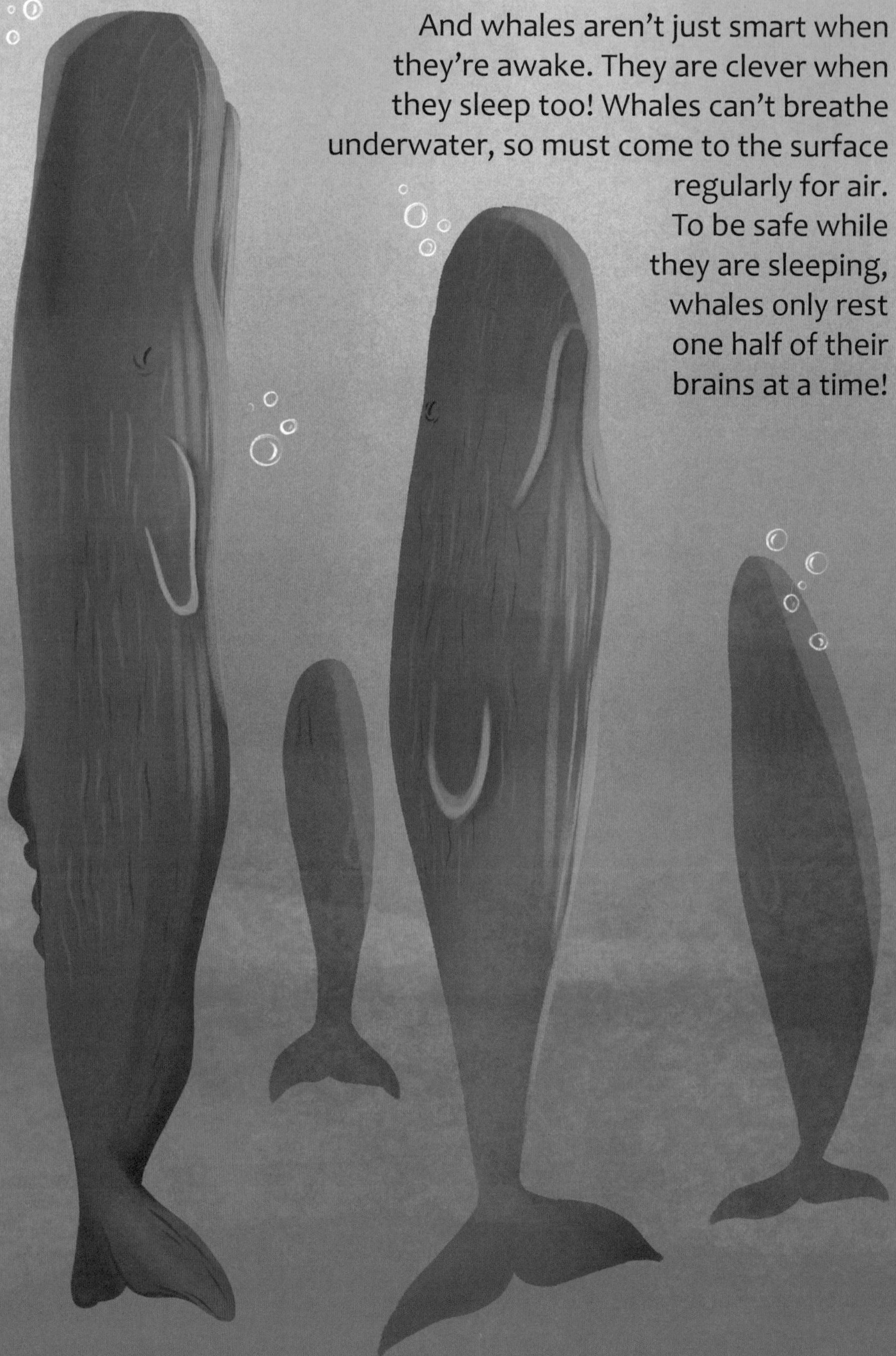

Sperm whales sleep pointing up
toward the surface, so they can
take in air quickly when they wake.

Intelligent animals like dolphins and whales are

VERY CURIOUS!

Humpback whales will sometimes approach boats, and can be seen **breaching** near them too. These large animals are curious about us, just like we are of them!

Studying whales, cetologists discovered something very important. A single whale can do as much good for our planet as **30,000 TREES!**

And it's because of their poop! Their **waste** feeds ocean plants and tiny creatures that release **oxygen** and store **carbon.** The oceans help make up to half of the air we breathe. So, it's important that we take care of them!

Whales and dolphins can do incredible things. They even work together! Whales sometimes swim next to dolphins to scare off **predators**. In return, dolphins help whales find food. Teaming up is one of the clever ways these animals survive.

Humans may be the smartest animals on land, but thanks to cetologists, we know that dolphins, whales, and porpoises are definitely the smartest animals in our oceans!

There are many species around the world, all in different shapes and sizes...

Record-breaking
WHALES AND DOLPHINS

Cetaceans may be the cleverest creatures in the sea, but cetologists have discovered lots of other impressive record-breaking features about them too...

The smallest dolphin is... HECTOR'S DOLPHIN!

This dolphin grows up to 5 feet (1.5 m) long. That's smaller than the average human! They are also the rarest of marine dolphins.

The biggest whale is... THE BLUE WHALE!

Not only is this the biggest whale, but the blue whale also holds the title for being the largest animal on Earth!

The deepest diver is...

CUVIER'S BEAKED WHALE!

This whale holds the title for the deepest diver of any cetacean. It held its breath for an impressive 222 minutes. That's just short of four hours without a breath!

The top predator is...

THE ORCA WHALE!

There is no denying this is the top predator of the cetaceans. It is at the top of its food chain, eating anything from fish to other dolphins and whales!

Cetologists are constantly learning new things. This information may change the more we learn about these wonderful creatures.

Wacky

WHALE FACTS

So we know whales and dolphins are pretty clever, but what else is there to know? What other facts have cetologists discovered about these smart underwater animals?

HUMPBACK WHALES ARE KNOWN FOR THEIR SINGING!

These whales are known for their complex and beautiful songs, which can last for hours. Whales use singing as a way to communicate and find each other.

NARWHALS HAVE LONG TUSKS THAT MAKE THEM LOOK LIKE UNICORNS!

A long time ago, some people believed that *narwhals* were underwater unicorns! The "horns" are actually **tusks** that grow from their mouths, and some narwhals even have two.

WHALES CAN LIVE FOR A VERY LONG TIME!

Bowhead whales have been recorded as living up to 200 years old. Many cetologists believe this is the longest living mammal!

WHALES HAVE FINGERPRINTS!

Humpback tails are as **unique** as human fingerprints. Cetologists can track the same whale its whole life just from noticing the markings on its skin!

THERE'S LOTS OF DIFFERENT TYPES OF WHALES!

So far, cetologists have discovered around 40 different species of whales in our oceans, each with their own stand-out features. But who knows, maybe there are more species to be found...

GLOSSARY

Blowholes – special holes on top of a whale or dolphin's head that it uses to breathe when it comes to the surface of the water.

Breaching – when a whale breaks through the surface of the water by jumping.

Carbon – one of the ingredients in carbon dioxide, a gas linked to rising temperatures on Earth.

Cetaceans – a group of mammals (see right) that include whales and dolphins. *Need help saying this? Look below!*

Environment – everything that is around us.

Habitats – the places where animals or plants live.

Mammals – warm-blooded animals (including humans) that produce milk to feed their young.

Migrate – when animals move from one place to another, usually to find food or because of changes in the weather.

Oxygen – an invisible gas in the air that plants produce, and people and animals need to breathe.

Predator – an animal that hunts other animals for food.

Sea sponges – soft, squishy creatures that live in the ocean and filter water to find food.

Species – a group of living things that share characteristics and features. For example, blue and beluga are different whale species.

Tusks – long, sharp pointed teeth.

Unique – something that is one of its kind; unlike all others.

Waste – things that are not needed anymore and are discarded, including poop!

HOW DO I SAY?

Beluga
buh-LOO-guh

Cetaceans
see-TAY-shuns

Cetologists
see-TOL-uh-jists

Cetology
see-TOL-uh-jee

Narwhals
NAHR-walls

THE BIG QUESTIONS ANSWERED

This is more than just a series of books; it is a complete resource. Accompanying each book is a variety of FREE material to engage curious kids with science.

www.thebigquestionsanswered.com

Use the QR code to visit the website, download free resources, and discover other books in the series.

On the website, find out incredible things about cetologists, including what they do, some of their greatest discoveries, and the people who have made a difference in this field of science.

The material is also available for home or classroom use, supporting all the information in this book.

Teachers' & Parents' Resources
With discussion prompts and questions, extra information, and facts around key topics.

Young Cetologists' Activity Pack
Fun activities for wannabe whale watchers, including creative writing, drawing, word searches, and much, much more.

BEETLE BOOKS

The Big Questions Answered is published by Beetle Books. Beetle Books is an imprint of Hungry Tomato Ltd.

First published in 2024 by Hungry Tomato Ltd
F15, Old Bakery Studios, Blewetts Wharf, Malpas Road, Truro, Cornwall, TR1 1QH, UK.

ISBN 9781835691335

A CIP catalog record for this book is available from the British Library.

With thanks to:
Editor: Millie Burdett
Editor: Holly Thornton
Senior Designer: Amy Harvey
Tim Cook for his valued contribution
The team at Beehive Illustration

Printed and bound in China.

Picture Credits:
(t = top, b = bottom, m = middle, l = left, r = right)
Shutterstock: Bekirevren 35tl; Catmando 34bl; Diana Askarova 33tr; G images.com 32-33b; Joost van Uffelelen 34mr; OSTILL is Franck Camhi 32mr; Sergey Uryadnikov 35mr; Solarseven 35bl.